Understanding Snapchat

Explore the definative guide to using snapchat.

Author: Benjamin Baker

© Copyright 2017 by Benjamin Baker - All rights reserved.

The follow eBook is reproduced below with the goal of providing information that is as accurate and reliable as possible. Regardless, purchasing this eBook can be seen as consent to the fact that both the publisher and the author of this book are in no way experts on the topics discussed within and that any recommendations or suggestions that are made herein are for entertainment purposes only. Professionals should be consulted as needed prior to undertaking any of the action endorsed herein.

This declaration is deemed fair and valid by both the American Bar Association and the Committee of Publishers Association and is legally binding throughout the United States.

Furthermore, the transmission, duplication or reproduction of any of the following work including specific information will be considered an illegal act irrespective of if it is done electronically or in print. This extends to creating a secondary or tertiary copy of the work or a recorded copy and is only allowed with express written consent from the Publisher. All additional right reserved.

The information in the following pages is broadly considered to be a truthful and accurate account of facts and as such any inattention, use or misuse of the information in question by the reader will render any resulting actions solely under their purview.

There are no scenarios in which the publisher or the original author of this work can be in any fashion deemed liable for any hardship or damages that may befall them after undertaking information described herein. Additionally, the information in the following pages is intended only for informational purposes and should thus be thought of as universal. As befitting its nature, it is presented without assurance regarding its prolonged validity or interim quality. Trademarks that are mentioned are done without written consent and can in no way be considered an endorsement from the trademark holder.

Table of contents:

Introduction 5

Chapter 1: Getting started 6

Chapter 2: How to use basic features on snapchat 11

Chapter 3: How to use the camera on snap chat 15

Chapter 4: How to use advanced features on snapchat 19

Chapter 5: Sending, viewing and chatting with snapchat. 23

Chapter 6: Phycological side of using snapchat 28

Conclusion 32

Introduction

Thank you for taking the time to download this book: Understanding Snapchat.

This book covers the topic of exactly how to download and use snapchat to its full potential, and will teach you not only how to use snapchat but the phycological reasons as to why everyone is so hooked on this specific application.

Snapchat is an instant image/messaging multimedia mobile application that was created by Evan Spiegel, Bobby Murphy and Reggie Brown. It was released to the app store in September 2011 accessible for IOS and Android phones. Here in this book you will learn how to download the app and use the functions of it in an interesting and helpful step by step process.

We will cover together; all the functions snapchat has to offer you to become efficient and tech savvy!

As an added bonus chapter I have included the 'Phycological side to using snapchat' to allow you to understand the reasons as to why people use this app and what its appeal is.

At the completion of this book you will have a good understanding of how snapchat works and be able to create an account and use it to its full potential for yourself.

Once again, thanks for downloading this book, I hope you find it to be helpful!

Chapter 1: Getting started

Snapchat is a well-known and well used application of this day and age. No matter your age let's do this together, it's time to get started.

To find this app you will need to access your computerized devices app store, whether it be an android phone, iPhone or iPad.

Click on the app store and then once you are in the app store look for the magnifying glass on your screen, (this is the search bar and is located at the bottom of the screen on iPhones). Now type in 'snapchat', then you will have several results show up but you need to click on the icon that has a fully yellow background with a white childish ghost in the center.

Here is the icon for those that don't know below.

Now once you have download the app you are ready to sign up and create a profile! Wooo! Click on the app that is now located on your screen then follow the procedure that snapchat has for creating a new account.

- Click the 'sign up' button at the bottom of the page

- Add your email address, if you do not have one you will need to create one at https://outlook.live.com then click continue.

- Create a username and click continue. This is what you will be known as by your friends so create your profile using your name so they can find you easily.

- Create a password and click continue. Make sure to create a password that you will remember.

- Add your phone number and click verify. You will now receive a text message on your mobile so keep it handy if you are doing this process on an iPad.

- Type the code that you have received as a message and click continue.

  Congratulations you have now successfully signed up to snapchat!

You will be prompted by snapchat in how to use the app so follow the procedure once again. This entire process should take you under 10 minutes.

Import friends from your device's contact list (optional):

Snapchat will scan your address book to look for other people who are using snapchat. You can skip this step if you wish by tapping '**Continue**' and then '**Not Allow**'. But you are on snapchat to send images/messages to your friends or gain a large friends base to make money so I suggest following the procedure and not skipping it.

Add friends:

You'll need people to chat with if you want to use snapchat. Adding someone allows you to send snaps to the person and view their public story, but the other person will need to add you back in order to reply to snaps and view a friends-only Story. So, for this part you will hope your friend knows it is you or you will have to let them know.

Many users don't allow snaps from users that they haven't added back. If you send a snap to someone who hasn't added you back, you will see "Pending" until your friend request is approved.

Add friends by username:

The quickest way to add a friend on Snapchat is to enter their Snapchat username. You can't add someone by email, real name, or phone number unless they're already in your Contacts and has enabled address book matching in their Snapchat settings.

Swipe down anywhere on the Snapchat camera screen.

- This opens your profile and options.

- Tap 'Add friends'. Doing so takes you to the Add Friends screen.

- Tap 'Add by username'. The username search bar will appear.

- Firstly know and Enter the username of the person you want to add. Start typing in the username and results will appear underneath the search field.

- Tap the "+" button next to a user that is your friend. Depending on their settings, you may see their profile image. Tapping "+" adds user to your friends list.

Profile pictures can look like this but with their face in the middle where the ghost is.

Add people by Snap code:

You can use Snap codes to quickly add people to your friends list.

- Take a photo or screenshot of someone's Snap code (profile picture that is shown above).

- Swipe down on the camera screen and click 'Add friends'.

- Click 'Add by snap code'.

- Select the image that contains the Snap code you want to add. The Snap code's owner will be added to your friends list (don't pick random people, only pick your friends).

Chapter 2: How to use basic features on snapchat

Once you first open snapchat you will have this picture on the left pop up but instead of a black screen you will have a view of whatever your camera is facing. The picture on the right is one of when you have taken a picture but we will go through that in the next chapter together.

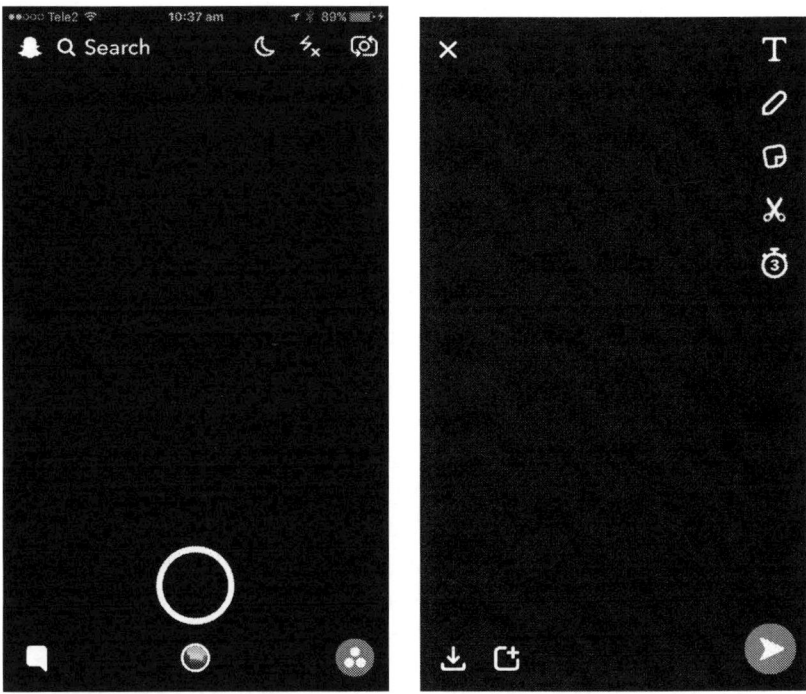

I will explain the icons on the left image now.

Snapchat (white ghost) icon:

If you click on the white snapchat ghost you will bring up your profile details. But you can also access this by simply swiping down on your screen with your finger.

In your profile details, you have the option to click

- 'Add me', it allows you to see who has added you.
- 'Add friends' gives you the opportunity to add your friends
- 'My friends' has the full list of your added friends.

Search icon:

This is the white magnifying glass followed by 'Search' it is a searching icon allowing you to search for your friends, to then add them.

Moon/sun icon:

This is the icon that is in the shape of a half moon or sun, it essentially tells you if your camera will be using its light or dark functions for day or night.

Lightning bolt with an X next to it icon:

This is a flash icon and you can choose to tap on it to turn the flash on or off depending on your current lighting conditions.

Camera icon:

This is the rectangle looking camera on the top right. Clicking on this will allow you to change the camera view from the back camera on your phone to the front camera. You use this function for taking a 'selfie' (a picture of yourself) or taking a regular picture of landscape or your friends.

Text box icon:

This is the white text box on the bottom left hand corner and if your friends have sent you a snap or just a message you will be notified here. It will show a number of messages you have waiting to see from 1-99. Depending on how many friends love you or how frequently you check your snap chat, will determine what number is near the text box.

Once you click on the text box it will bring up all of the recent snaps from your friends, there are 20 types of shapes that could potentially show.

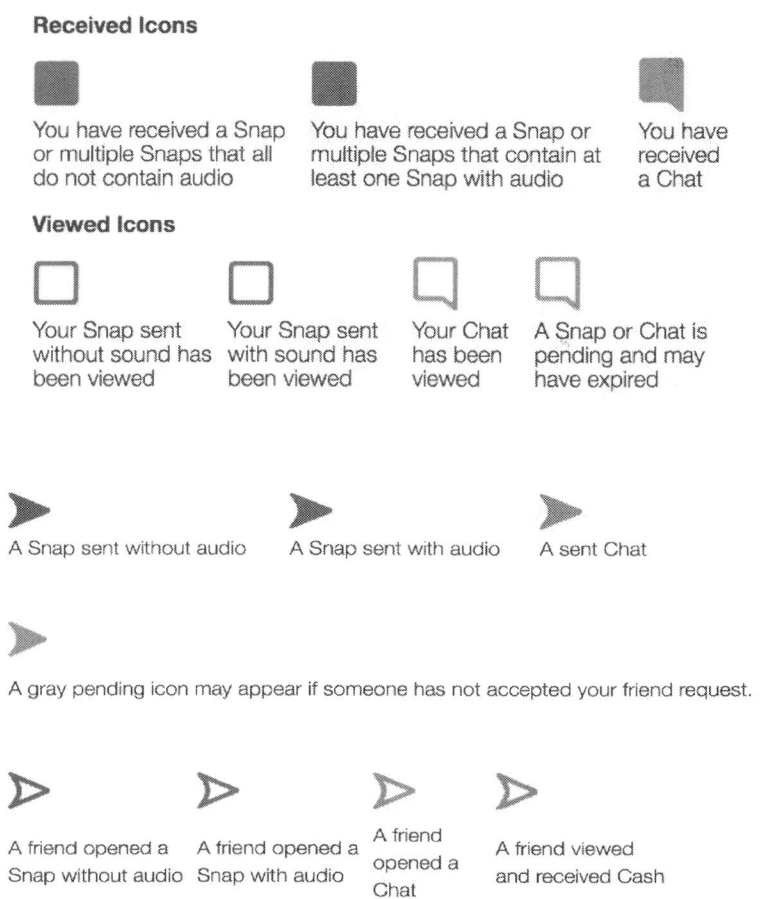

A screenshot has been taken of your Snap without audio   A screenshot has been taken of your Snap with audio   A screenshot has been taken of your Chat

Your Snap sent without sound has been replayed   Your Snap sent with sound has been replayed

Singular little circle:

The singular little circle is located below the big one at the bottom of the screen and it is essentially a quick link for you to view your recently posted snap chats or device cameras images. You can then choose to select one of the pictures from that device and post it for your friends to see, or you can repost a picture that you once already did. If you want to show someone or everyone that picture again.

Purple circle icon:

This icon is in the bottom right hand corner and it has three mini circles inside one purple circle. This icon gives you the ability to view your story or your friend's stories.

Chapter 3: How to use the camera on snap chat

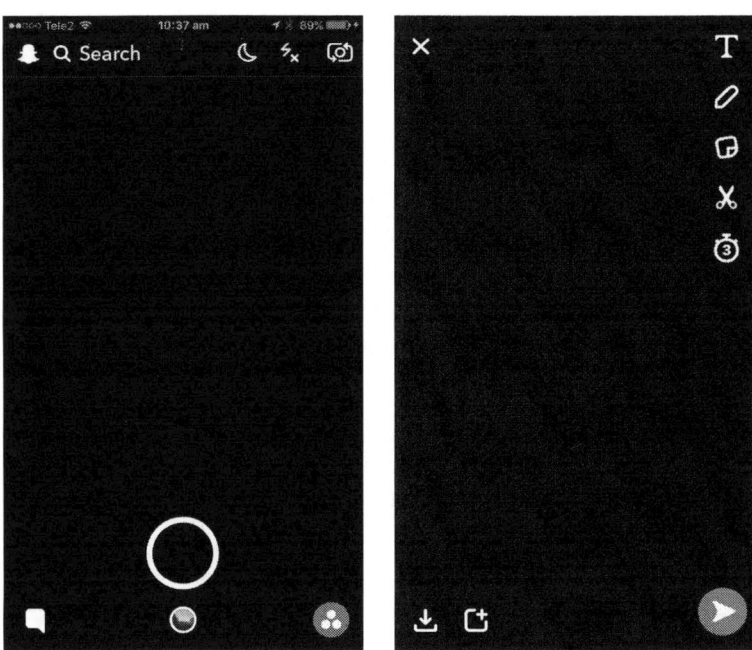

Now for creating a snap of your life around you or of that sexy face of yours. This is the main reason as to why snap chat is used. It is to show your friends what you are doing in your life at that moment or get self-satisfaction of them saying you look beautiful. We are all only human and we love to hear that we look good because that makes us feel really good inside.

Decide the camera's direction:

If your device has a front and a rear camera, you can switch between them by tapping the button with two arrows forming a rectangle in the upper-right corner of the Snapchat camera screen. (As shown earlier)

You can also double-tap the screen anywhere to toggle between front and rear-facing cameras.

Take the picture:

Tap the larger circle button at the bottom of the screen to take a still image Snap. The camera will capture whatever is currently displayed on the screen at that particular moment.

Record a video:

Press and hold the circle to record a video Snap.

You can record up to 10 seconds of video in single snap. Snapchat will only record as long as you are holding the button. If you were to let go of the circle before the 10 seconds is up it will instantly stop recoding. Although if you continue to hold the button it will not do any extra footage than the designated 10 seconds.

Bad image or video?:

If you don't like the picture or video snap you took, once you have taken the shot this button in the upper-left corner deletes it 'X'. This feature enables you to continuously take pictures or videos until you are happy with it. Remember natural beauty is the best beauty, you don't need to fake it.

Letter 'T' icon:

The 'T' icon is for text; this feature allows you to write up to 6 lines of text. You can write whatever you please and this writing will be placed in front of your image. The 6 lines are a new feature as when this app was first created you only had the capability to write one line of text. As well as the new 6 line feature you can now change the color of your text. Once you click on the 'T' you will notice a color line chart on the right of your screen, slide this up and down to change the color that best suits your image.

You can also get into the writing text icon by simply taping anywhere on your screen once you have taken the image or video.

Then once you have written your text, to can also move its position. You can do this by simply placing your finger on the screen and dragging it up or down. Once you like the position of the text release your finger.

Snap timer icon:

Tap or click the timer button (right hand side, bottom icon that looks like a stop watch) to limit the duration of photo Snaps. Photo Snaps display a stopwatch button that allows you to set a duration between 1 and 10 seconds. This is the amount of time a Snap will be displayed before it disappears.

Video Snaps do not have timers. They will play through the length of the recording and then close.

The drawing icon:

This is a new and improved feature that is located on the right-hand side of your photo and in the shape of a pencil (below the 'T'). Click on the pencil and a little hand starts flashing this is showing you that you can make your pencil line thicker or thinner by pinching in or out. Now Drag your finger to draw freehand images on your Snap. The slider on the right will allow you to change the color of the line. You can also tap the Undo button to erase the last drawing you made in case you don't like the last line you drew. Its located on the top of your screen to the left of the pencil (a white arrow forming a circle)

Then there is another new feature that is located below the color bar, you can click on the 'red heart' and it shows a list of different emoji's. Select one of the emoji's and tap the screen once to put one image down or tap, hold and drag your finger along the screen to place several of them down in a row.

Now you know the art of drawing with snap chat let your master pieces come to life!

Chapter 4: How to use advanced features on snapchat

Now that you have a basic level of understanding on how to navigate your way around snapchat let's get into a few fun advanced features that snapchat has.

The lens changing feature:

If you tap anywhere on the camera screen, you can use the Lenses feature in snapchat to scan your face and apply crazy effects. This can be done with either the front or the rear camera, but you'll find it easier if you're taking a selfie. Remember to tap the camera button in the top right corner to switch to your front-facing camera.

Keep in mind, you **can't** apply the lenses effects to snaps you've already taken. You'll need to activate lenses before you take a snap.

Swipe left to preview lenses effects. Many offer prompts so be sure to "Open your mouth" to ensure you get the best of the effect's.

The selections of effects rotate on a daily basis, with the oldest Lens in the list being replaced by a new one. If you don't see the effect you want, check back in the next few days.

If you can't get Lenses to start, your device may not be compatible with snap chat's requirements. Lenses requires Android 4.3 or newer, or the iPhone 4S or newer. Note that older iOS and Android devices may not work, even if they're running the latest software versions.

Now take or record a Snap with your chosen effect. While the effect is active, you can take and record Snaps just like you would normally. Click the circle button (which displays the lens you chose) to take a photo Snap with the current effect, or press and hold to record a video.

Face swapping:

Open the Lenses feature. Follow the steps in the previous section to open the list of available Lenses.

Select the "Face Swap" lens. Swipe through the menu until you find the Face Swap option. The icon has two happy faces on a yellow background, with arrows pointing toward one another.

If you don't see Face Swap, it may have rotated out of the available selection for today. Check back in a few days to see if it's back.

Now put your face and your friends face in the two circles provided these will turn yellow when a face is aligned with it. Once both faces are aligned, they'll be swapped! You don't necessarily have to be standing next to each other, you just need to have both faces aligned in the frame.

Take a Snap. Just as the other lenses, you can take a Snap by tapping or holding the circle at the bottom of the screen.

On the bottom, left hand corner there is an arrow to save your snap. If you want to be able to show the picture to your friend or family member that isn't using Snapchat, save it to your phone or to Snapchat Memories by tapping the **Save** button.

The Sticker icon:

This icon is located below the 'Draw' icon. The sticker icon looks like a post it note. Tap on the sticker button to see stickers, emoji's and bitmojis. You can place these anywhere on your picture, and as many as you please.

Swipe left and right on the sticker screen to see different categories. There are several different stickers to choose from so don't hold back.

Now click on a sticker to add it to your snap, you can drag it around the screen with your finger to place it where you'd like.

Size of the sticker:

Use two fingers to adjust the size and rotation of sticker. Pinch the sticker in and out to adjust the size. Rotate your two fingers on the screen to rotate the sticker.

To make a sticker:

Tap the scissors icon at the top of the screen, then use your finger to outline any part of the image/video, such as a person's face and let go. Now you've created a sticker that will be saved and allow you to move/put it anywhere on the screen with this or other snaps.

Chapter 5: Sending, viewing and chatting with snapchat.

Sending a Snap:

Once you're happy with how your snap looks, tap the blue **Send** button in the lower-right hand corner to open your snapchat contacts list. Then once that is open select who you want to send it to out of your contact list by tapping on their name. Followed by the send button.

To save a snap:

Click the "Save" button to save the snap to your phone. It's in the lower-left corner and looks like a downward-pointing arrow in a box. When you send a snap, it's gone forever, so if you would like to keep it then make sure to save it.

You can send the snap to as many people on your list as you like. Your friend or family member can view your snap once before it disappears, unless they use their 'replay' on it (we will go through replays more later).

My Story:

Snaps that are added to your story can be viewed by anyone who has access to your story for the next 24 hours. After 24 hours, the snap is permanently deleted. By default, anyone can view your story as long as they know your Snapchat username. But it is highly unlikely for a random person to guess your user name and look at your story. The snaps that are in your story can be replayed as many times as the viewer wants for up to 24 hours, so if your friend really likes that picture you took they might look at it a couple of times. Or if it is an extremely funny one they may show whoever is around them.

You can change your story's privacy settings by swiping left on the camera screen and tapping the settings button '□' in the upper-right corner of the screen.

Viewing Snaps and Stories

Click on the 'Chat' button to see snaps you received. It's the speech bubble square in the lower-left corner of the screen. You'll see a number in this button indicating how many new Snaps you've received, if there is more than one to view.

- Red Snaps are photos.

- Purple Snaps are videos.

- Blue Snaps are text chats.

Tap the Snap that you would like to view. The Snap will disappear as soon as the timer runs out. But sometimes if you're in a bad reception area it may take a while to load, in these cases be patient.

If you've received multiple snaps from the same person, you'll be able to view them back-to-back, it will show a timer of how long you can continue to view this in the top right-hand corner. You can tap with a second finger anywhere on the screen to advance to the next snap early if you don't want to view them all for the full-time period.

You can take screenshots of snaps, but the sender will be notified that you did this, remember when we looked at all of the different kind of shapes next to a snap picture. One of those shapes show when someone has taken a screen shot. Most

Snapchat users don't like it when people take screenshots, as it's contrary to the spirit of Snapchat. Check with the sender (your friend) for permission before screenshotting, or you may get blocked from viewing their snaps.

To reply:

Tap and hold a snap again to use your single-use Replay after you have viewed it once. Snapchat allows you to replay a Snap once before you navigate away from the chat screen. If you move off the screen, you lose your instant replay.

Snapchat stories:

Swipe left on the camera screen to open Stories. By doing this, takes you to the screen where you can view your friends' Stories or you're own.

Simply tap on your friend's story if you wish to view it. When you tap a Story, it will begin to play from the oldest Snap. Each Snap will play through its timer and then move on to the next one. If you tap on the right side of the screen while the story is playing it moves on to the next Snap, or the left side to go back one Snap.

Swipe down on the screen to quickly close a story if you have had enough or it is inappropriate for some reason.

Replying to a Snap in a Story:

At any time during a story, you can send a reply to the Snap you are viewing. Swipe up from the bottom of the screen to open the onscreen keyboard. Then start typing what you wanted to say and click send.

My Story:

This will only appear if you've added Snaps to your Story. But all you have to do is tap it once and it will play all of your snaps. Tap the ^ button at the bottom of your Story to see details. You'll be able to see how many people have viewed each snap, as well as how many have screenshotted it. You can tap the Trash button to delete that snap from your story if you didn't like it, or tap the download button to download it on to your phone to keep.

Chatting with Snapchat

Tap the Chat button: It's in the lower-left corner or swipe right on the camera screen to open your messages.

Tap the "New Chat" button in the top right corner. This opens a list of all your Snapchat contacts. Then tap the user/friend you want to chat with. This will open the chat screen.

Now a keyboard will pop up and you have the opportunity to type whatever you please to that person, just like you would do so in a normal message.

Use the buttons above the keyboard to enhance your messages.

This is a new and improved feature on snapchat, it is almost as effective if not more effective than other social media networks in regards to a constant talking app.

- Tap the "Photos" button to open your phone's pictures. Doing so allows you to send pictures that have been saved to your phone or iPad.

- Tap the "Phone" button to place an audio call. The recipient will be notified that they are receiving a call. (This app has developed so much over the last 6 months).

- Tap and hold the "Phone" button to record an audio note. You can record up to ten seconds of audio, just like a normal video that the recipient can listen to when they are in the chat.

- If you want a video chat, tap the "Video" button to place a video call. The recipient will be notified that they are receiving a video call from you and can decide to answer or not.

- Press and hold the "Video" button to record a video note. Like the audio note, you can record up to ten seconds of video that your friend can view.

- If you want to add stickers, tap the "Happy face" button to open stickers, emojis, and Bitmojis. Chat has a greater amount of sticker options available than regular Snaps. Scroll down with your finger through the list to view all of the available stickers.

- Now finalise that creation by tapping the 'send button. Then we just hope your friend enjoys the message and replies to keep the conisation flowing.

Chapter 6: Phycological side of using snapchat

Snapchat is used amongst a lot of people because it is an intimate way of chatting to someone. Researchers from the Cornell Social Media lab interviewed 25 students (8 male and 17 female) and asked them about how they use Snapchat. There most common response was that since snaps in Snapchat disappear after they are viewed, the app is "a lot less formal" than Facebook and other social networks. For that reason they love it because they feel a lot more relaxed sending whatever they want to their friends.

Alongside with this statement it was a common trend that people were saying that unlike Facebook, you have less friends on here. This means you will only view snaps from those that are close to you. By having only those that are close to you as a friend you don't mind what people think of your snaps as much. This leads to more comfortability knowing that you can take a snap and that person will only get to view it for a few seconds and that is all.

Snapchat is an app that allows you to live in the moment rather than living in the past and looking at images 2 weeks ago. It gives you the opportunity to view your friends lives without constantly messaging them. You can view what they did in the past 24 hours without actually contacting them. Which keeps you close and in the loop of their life.

Some of the students also stated that it's because of these factors that they feel more comfortable in expressing exactly how they feel with an image that will not be judged. On Facebook, Twitter or Instagram once you put it up for everyone to view, people comment and judge it which can lead to a lot of people holding back and only posting good images. Rather than truly emotional images that show exactly how you feel in the moment.

People seek the need to engage informally with their best friends and snapchat provides this. We all want to have that feeling of self-worth and importance in somebody's life. By sending snaps and seeing how many you receive in one day, gives us that satisfaction of importance and self-worth. Sometimes we are not able to see that person in real life or say/do the things we truly want to do when in front of that person. Which then perhaps leads to the intimacy side of the app, it not only gives you an outlet for you to show your emotions to someone through an image that they cannot keep. But is gives you the opportunity to flirt with your body, building and crossing relationship boundaries.

Nudes:

People sometimes use this app to get inner security/ satisfaction by sending another person they are interested in a 'nude' (a naked image of the body). This not only provokes compliments that would not normally be said in real life but it can lead to sexual desire and want for that person. Snapchat can build a new relationship faster than what you would in real life, even if you are an extremely shy person. This is all because of the freedom of the photo being permanently deleted in a matter of 10 seconds or less. (unless someone screenshots it, in which you can see if they do)

When you send a sexual image to someone your heart will race quite fast because it is a very private/intimate part of your body and you are essentially teasing the other recipient.

Let's go through a deep feeling, imaging you are lying in your soft comfortable bed with the door locked and you have just returned from a third date with someone special...

Your mind is racing thinking of those new loving, happy thoughts about that beautiful man/women you just spent time with. You know instead of coming home after a nice candle lit dinner at 11pm on a Friday night, the last thing you want is to be lying in bed alone. So instead of lying in bed attempting (but failing) to fall asleep, you send a picture on snapchat.

In this picture, you have no top on and you send a sexually luring image of your top half (only teasing) saying 'thank you for a nice night I had so much fun'. Your body is feeling excited more than ever right now awaiting a reply, you watch the picture you sent refreshing the page every 10 seconds to see if the other person has seen it yet or not. Then the solid blue box turns into a outlined blue box, now you're thinking great they have seen it!!! (maybe thinking, oh no did I go too far? Does this person like my body?)

Then you receive an image back, you tap it... and get a reply back with an equally as sexual image!!! You now love what your eyes are seeing for that 10 second period and you want more, you send another sexually attractive image but not your top half, maybe a sneak peak of somewhere else on your body... you await another reply and you feel like your sexual desire is rising more and more every second waiting in suspense. Then after a minute or two of waiting you get what you believe to be as a mind-blowing image... an image of their cat hahahaha I'm kidding we are not going to get extremely sexual here I will leave your imagination to do the work.

This particular feeling is what we are all after when searching for a new lover (the tease). It's what makes being newly interested in someone so much fun, you have the opportunity to fulfill your sexual desires with someone without being too intimate and giving away everything from the word 'GO'.

ONLY SEND PICTURES YOU WILL NOT REGRET, JUST INCASE SOMEONE DECIDES TO KEEP IT FOREVER.

Conclusion

Thanks again for taking the time to download this book!

All in all, you will find that Snapchat is a great visual app to keep in contact with your closest friends. It's a visual "poke" or a way of keeping in touch, satisfying the needs for connection and belonging while maintaining privacy and freedom.

You should now have a good understanding of how snapchat works, and be able to use snapchat to its full potential.

If you enjoyed this book, please take the time to leave me a positive review on Amazon. I appreciate your honest feedback, and it really helps me to continue producing high quality books.

Made in the USA
Lexington, KY
02 August 2017